3 1994 01269 7378

0/05

SANTA ANA PUBLIC LIBRARY
NEWHOPE BRANCH

AR PTS: 1.0

EARTH EXPLORATIONS

Sedimentary Rocks

JENNY KARPELENIA

J 552.5 KAR
Karpelenia, Jenny
Sedimentary rocks

$18.50
NEWHOPE 31994012697378

PERFECTION LEARNING®

Editorial Director: Susan C. Thies
Editor: Mary L. Bush
Design Director: Randy Messer
Book Design: Tobi S. Cunningham
Cover Design: Michael A. Aspengren

A special thanks to the following for his scientific review of the book:
Kristin Mandsager, Instructor of Physics and Astronomy,
North Iowa Area Community College

Image Credits:
©Lloyd Cuff/CORBIS: p. 20; ©Phil Schermeister/CORBIS: p. 22 (bottom); ©Richard Cummins/CORBIS: p. 33; ©Archivo Iconografico, S.A./CORBIS: p. 39

Photos.com: pp. 3, 6, 7, 8, 9, 11, 12, 14, 15 (bottom), 16, 17, 19, 21, 23 (bottom), 24 (bottom), 25, 26, 27 (bottom), 29, 31, 32 (bottom), 34, 36 (top), 40, 41, 42, 47, 48; Corel: pp. 5, 23 (top), 24 (top), 27 (top), 37, 45, 46; PLC images: 10, 13, 15 (top), 22 (top), 28, 32 (top), 36 (bottom), 38, 43

Text © 2005 by Perfection Learning® Corporation.
All rights reserved. No part of this book may be reproduced, stored in a retrieval system, or transmitted in any form or by any means, electronic, mechanical, photocopying, recording, or otherwise, without prior permission of the publisher.
Printed in the United States of America.

For information, contact
Perfection Learning® Corporation
1000 North Second Avenue, P.O. Box 500
Logan, Iowa 51546-0500.
Phone: 1-800-831-4190
Fax: 1-800-543-2745
perfectionlearning.com

1 2 3 4 5 6 PP 09 08 07 06 05 04

ISBN 0-7891-6233-4

Table of Contents

1

A Rockin' Vacation

You're on vacation with your family. Your patience is wearing thin. Your sister is driving you insane, and your parents' choice of music is ancient. Everyone is exhausted. The drive through the desert is dusty and hot. All you can think about is a cold drink and maybe a burger and fries. Will you ever get there?

The Grand Canyon

After what seems like an eternity, your dad finally announces, "We're here!" Where is here? You have arrived at the Grand Canyon in northern Arizona. This natural wonder is a site that draws millions of tourists every year. But what's so special about it? It's just made of plain old **rock**. But over billions of years, this "ordinary" rock has become an awesome sight to see.

Let's Get Rockin'

What is a rock? A rock is a combination of two or more **minerals**. Minerals are nonliving substances made up of one or more **elements**. Elements are nonliving substances made of only one type of **atom**, or particle. For example, silicon and oxygen are elements that make up the mineral quartz. Quartz cemented together with sand grains makes the rock quartz sandstone.

silicon + oxygen = quartz (mineral)
quartz + sand grains = quartz sandstone (rock)

The Grand Canyon began as a mountain range billions of years ago. Over time, **weathering** and **erosion** wore down these mountains. When water covered the eroded mountains and then receded or dried up, layers of **sediment** were left behind. These layers of rock pieces, minerals, shells, and other particles were squeezed together into layers of **sedimentary** rock that piled up on one another. This cycle continued over millions and millions of years. When the Colorado River began to flow through the area, it carved deep canyons through the layers of rock.

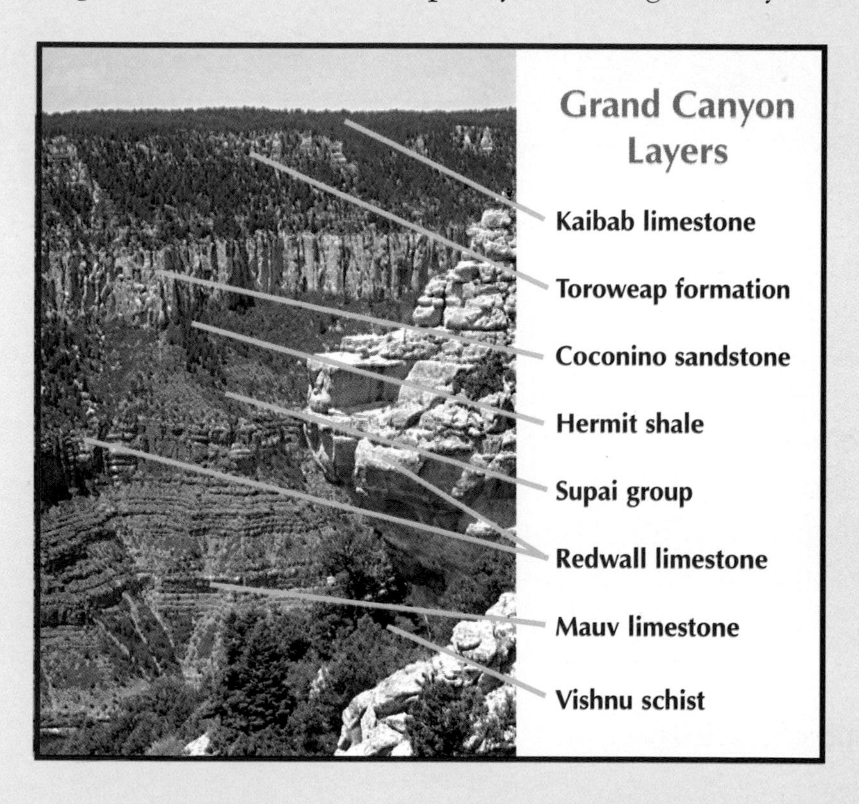

During your stay at the Grand Canyon, you learn about the different layers of rock that make up the canyon. The top layer of rock, the kaibab limestone, is about 250 million years old. Any newer rock that may have once been on top has now been eroded away by water and wind. Many layers of sandstone, shale, and limestone alternate below the top layer.

Colorado River

You hike some trails and ride a mule down to the bottom of the canyon. Layers of schist, gneiss, and granite are found here. These rocks are almost two billion years old. Here you can see the Colorado River cutting through the canyon. Its powerful flowing water has carved through the towering rock, exposing the different-colored layers. The rushing water has carried away tons of rock pieces. It **deposits** the pieces downstream, some as far away as the Pacific Ocean along the California coast.

After a few days exploring the Grand Canyon, you decide that maybe rocks can be interesting after all.

Carlsbad Caverns

After your time at the Grand Canyon, your family hops back in the car and heads to New Mexico. The next stop on your family vacation also involves rocks, but these rocks are underground. Your destination is Carlsbad Caverns.

Carlsbad Caverns were also formed by water and sediment. Over 250 million years ago, the Carlsbad area was covered with a shallow sea. When sea creatures died, their remains formed a 400-mile-long limestone **reef** in the shape of a horseshoe. Eventually the sea dried up and the reef was buried under minerals such as salt and gypsum. Then just a few million years ago, the buried reef was pushed upward by the movement of the Earth's crust. The reef was exposed when the minerals and other sediment that had covered it were eroded away. As rain trickled down into cracks in the limestone, the acid in the rain **dissolved** and reshaped the rock. The underground caverns, or caves, were created. Carlsbad Caverns National Park contains more than 80 of these caves.

Stalactite formations hang from the ceiling of Carlsbad Caverns.

A Rocky Reef

Limestone is a sedimentary rock formed from the remains of shelled sea creatures, such as snails, oysters, and clams. When layers of shells are pressed together over time, limestone rock is created.

While visiting Carlsbad Caverns, you tour rooms and passageways carved out of the limestone. You see interesting formations inside the cave. Some look like soda straws and icicles hanging from the ceiling. Others are spikes growing up from the floor. Some even look like popcorn and Swiss cheese. The tour guide explains how these unique features were formed when rainwater dissolved the limestone and left behind small amounts of **calcite**.

All of this cave exploration has made you hungry. It's a good thing that your family is able to enjoy lunch at a cafeteria located 75 stories underground. While eating, you discuss your next family vacation. Perhaps you'll visit Mount Rushmore, the famous presidential statue carved in a mountain of rock, or Hagerman Fossil Beds National Monument, a huge mass of plant and animal fossils more than 3.5 million years old.

Mount Rushmore

Now you can't wait to get home and tell your friends about your visit to the Grand Canyon and Carlsbad Caverns. Even though your trip got off to a rocky start, it turned out to be a rockin' vacation!

2

It's All in the Sediment!

Most of the rocks you saw at the Grand Canyon and Carlsbad Caverns were sedimentary rocks. Seventy-five percent of the Earth's surface is covered with these rocks. So what are sedimentary rocks? And what is all this sediment everyone keeps talking about?

Falling to Pieces

Sediment is small broken pieces of rocks, minerals, sand, soil, shells, and other materials that make up larger rocks. Most sediment is formed through the process of weathering. Wind, water, and ice are the three main forces of weathering.

Wind carries tiny sand grains and other particles and pelts them into larger rocks. The grains slowly wear away the rocks' surfaces. Eventually wind can actually carve right through softer rock, leaving natural arches, bridges, and large pillar-shaped structures called *buttes* (byoots).

Water can have many weathering effects on rocks. Water moving in rivers, lakes, and oceans has the power to wear down larger rocks into smaller pieces. Fast-moving rivers can cut deep valleys through rock. Rocks tossed around by waves in lakes and oceans smash into one another. Sharp, jagged edges break off into sediment while the remaining rocks are smoothed into smaller, rounder stones. Ocean waves also beat upon the shore, making the rocks there smaller and smoother.

TRY THIS!

See how the speed of moving water affects weathering. To do this, you need two pieces of unwrapped hard candy (same kind and size), two clear jars with lids (same size), and a measuring cup. Pour one cup of water into each jar. Drop a piece of candy into each jar and put the lid on. Put one jar in a safe place, and let it sit for a day without moving it. Keep the other jar nearby and shake it throughout the day. At the end of the day, compare the pieces of candy. What do you notice?

You should see that the candy in the shaken jar has dissolved or been broken up more than the one in the still jar. This is because the quick movement of the water in the shaken jar causes the candy to weather faster, just as rocks weather more quickly in fast-moving water.

Heavy downpours or floodwaters can also rush across the ground and wash away rocks. Rainwater mixed with the carbon dioxide gas in the air forms carbonic acid. This acid can eat away at rocks on or under the Earth's surface.

When water flows into cracks in rocks and then freezes, ice becomes a weathering factor. Ice takes up more space than water. It expands in the cracks and pushes on the rock. With enough pressure, the ice will split the rock into pieces.

Glaciers are huge chunks of slow-moving ice that scrape and carve rocks in their path. Glaciers can also polish underlying rocks.

All of these forces of weathering cause rocks to fall to pieces, or sediment.

A Few More Forces

Plants, animals, and chemicals can also weather rocks. Tree roots growing in rock cracks can split rocks apart. Animals break up rocks on the Earth's surface when they dig holes in the ground. Lichens are living things that grow on rocks. They release a chemical that causes the rocks to crumble. Chemicals in the air and water can damage buildings and statues carved out of rock and wear down underground rock.

Sediments on the Move

What happens to the pieces once they're broken? Erosion takes over. Erosion is the process of moving sediment. Gravity, wind, water, and ice erode, or move, sediment.

Gravity pulls sediment downward. Pieces of rocks that break off on hills or mountains tumble down the slopes toward lower ground. Wind picks up lighter sediment and blows it to new locations. Rivers carry sediment downstream and deposit it near the river's mouth. Rainwater washes sediment across the ground to new destinations. Glaciers slowly carry sediment from one area to another. All of these forces work together to move sediment around until it settles down to make new sedimentary rock.

Becoming a Rock

So rocks fall to pieces and the pieces, or sediment, get moved around. But how does the sediment become a sedimentary rock?

When gravity, wind, water, or glaciers deposit sediment, it forms layers. The bottom layer is the oldest layer, meaning the sediment has been there the longest. New layers of sediment are deposited on top of one another, forming a "stack" of sediment. The top layer on the stack is the youngest, or most recent, layer.

Over time, the layers of sediment are compacted, or squeezed together, due to the weight of the other layers. The sediment is packed tighter together, leaving smaller spaces between particles. The compacted layers are then **cemented** together. Water containing minerals fills in the spaces between the sediments. As the water **evaporates**, the minerals **precipitate**, or settle, out of the water. These minerals act like a glue cementing all the sediment together. Eventually the minerals and sediment become permanently locked together, forming solid rock.

Sedimentary rock can also be formed when water evaporates from shallow bodies of water. Minerals in the water are left behind and settle on the bottom. As they harden, the mineral **crystals** form sedimentary rock.

Once a Sedimentary Rock, Not Always a Sedimentary Rock

Sedimentary rocks are just one type of rock. There are two other types of rocks—igneous and metamorphic. Igneous rocks are formed by the cooling of magma, or liquid rock, at or near the surface of the Earth. Metamorphic rocks are sedimentary or igneous rocks that have been changed by heat and/or pressure. All three types of rocks are connected in the rock cycle. Metamorphic and igneous rocks are worn down into sediment, which becomes sedimentary rock. Changes in temperature and pressure cause the sedimentary rock to change into metamorphic rock. When pushed underground, this rock melts and forms new igneous rock. In time, this rock is pushed to the surface, where it is broken down into sediment and begins the cycle all over again.

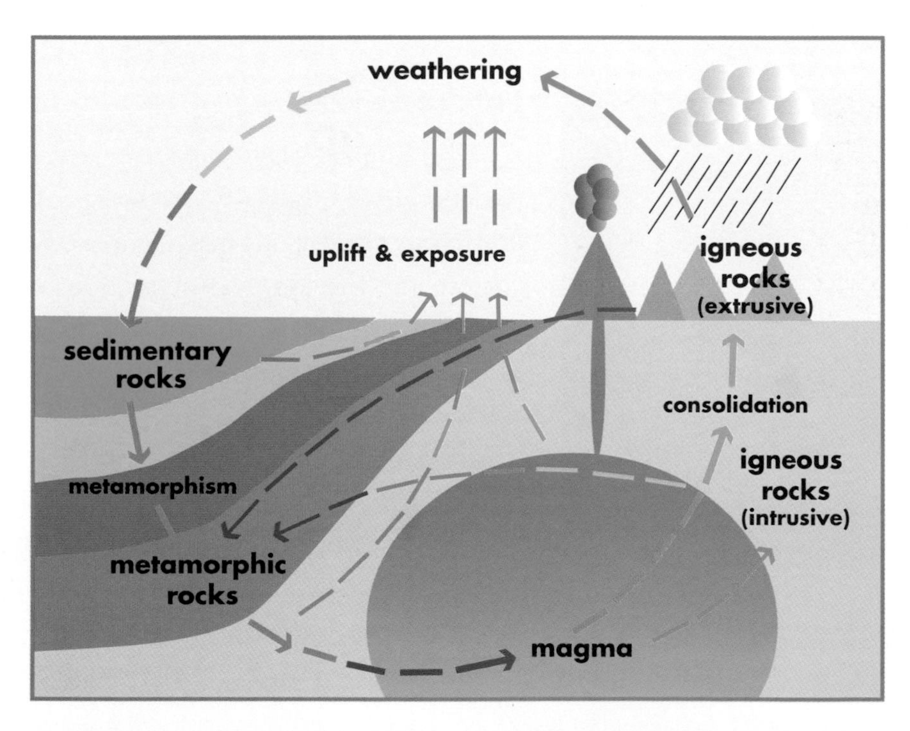

3

Looking for Clues

How can you tell if a rock is a sedimentary rock? Look at the rock's features. They give you clues as to how the rock was formed. Sedimentary rocks have special characteristics, such as layers, fragments of other rocks or shells, sediment markings, and **fossils**.

Lots of Layers

The biggest clue to sedimentary rocks is their layers. These layers are also known as **strata** or beds. Strata make it easy to see the beginning and end of a particular sediment **deposit**. Each time sediment is left behind in an area, it settles in a flat bed, or layer. The next layer of sediment will settle on top of that one. The change in color and/or texture of the layers will show where one layer ends and the next one begins. For example, a layer of white limestone may be followed by a layer of brown sandstone, which may be followed by a layer of gray shale.

TRY THIS!

Create your own strata model. Choose five pieces of different-colored breads (white, wheat, rye, pumpernickel, etc.). Stack the pieces on top of one another, alternating colors. Place several heavy books on top of the stack. After several hours, remove the books and observe your sedimentary rock strata.

Finding Fragments

Fragments, or pieces, of rocks, minerals, or shells within another rock are usually a sign that it is a sedimentary rock. Since sediment is broken pieces of rocks, minerals, shells, and other materials, it makes sense that some of these fragments would be visible in sedimentary rocks. Rock or mineral fragments of all different sizes can be cemented in sedimentary rock. Whole shells or pieces of shells can also be trapped in sedimentary rock.

Making Their Marks

A variety of special marks can be spotted on sedimentary rocks. Ripple marks are wavy ridges in a rock. They are formed from shallow ocean water waving back and forth on the sand underneath. When the sand hardens into sandstone, the ripple marks are still visible. Wind can also create ripple marks in sedimentary rocks.

Sedimentary rocks can also have rill marks, raindrop impressions, and mud cracks. Rill marks are small indentations made when water moves slowly over sediment in brooks or streams. Raindrops can also leave small dents on sedimentary rock. Tiny mud cracks can appear when mud dries and hardens into sedimentary rock.

Not all sedimentary rocks will have these markings. But rocks that do have one or more of these impressions are likely sedimentary rocks.

Sandstone with ripple marks

Hints from Fossils

Sometimes plant and animal fossils become trapped in sedimentary layers. When animals die in or near water, sediment often buries the bodies. Softer body parts rot away. Tougher parts, such as bones, teeth, and shells, absorb minerals from the sediment and become very hard. These fossils then become part of the rock that the sediment forms.

Plants can also become fossilized. Leaves and flowers leave imprints when the mud around them hardens into rock. These prints are then stuck in the layers of sedimentary rock.

Fossils of trilobites, which are extinct ocean animals with a hard outer shell

Sorting the Sediment

Once you have a clue that a rock is indeed sedimentary, you can classify it as one of three types of sedimentary rocks—**clastic**, **organic**, or **chemical**. Each kind of rock is made of different sediment. Clastic sedimentary rocks are formed from broken fragments of **preexisting** rocks, minerals, and shells. Organic sedimentary rocks are made from the remains of plants and animals. Chemical sedimentary rocks are formed when dissolved minerals settle (precipitate) out of water. You can explore these three groups of sedimentary rocks further in the following chapters.

4

Clastic Rocks

It's All About Size

Clastic means "broken." Clastic sedimentary rocks are made from the broken fragments of preexisting rocks, minerals, and shells. These fragments vary in size. Geologists use their eyes, magnifying glasses, and even rulers to measure **grain** size. They also use their fingers, mouth, and teeth to feel the size and texture of grains. Clastic rocks are classified according to the grain size of their fragments.

A Piece of Information

A clast is a broken piece of rock.

A Rockin' Career

A geologist is a scientist who studies the structure and history of rocks, minerals, and soils on Earth.

Large Grains

Boulders, cobbles, and pebbles are the three largest grain sizes. Boulders are larger than 10 inches across. Cobbles range from 2.5 to 10 inches. Pebbles measure from .08 to 2.5 inches in width. Gravel is a mix of grain sizes between .08 and 10 inches. Gravel consists of rock chunks mixed with sand.

Conglomerate is a clastic rock that is a combination of gravel, pebbles, and/or cobbles cemented together. It is sometimes called *puddingstone* as it can look like lumpy pudding. The fragments in conglomerate rocks are smoothed and rounded from the movement of water around them. They usually travel quite a distance in a river, lake, or stream before being deposited as sediment.

Lumps of Rock

The word *conglomerate* comes from a Latin word meaning "lumped or rolled together."

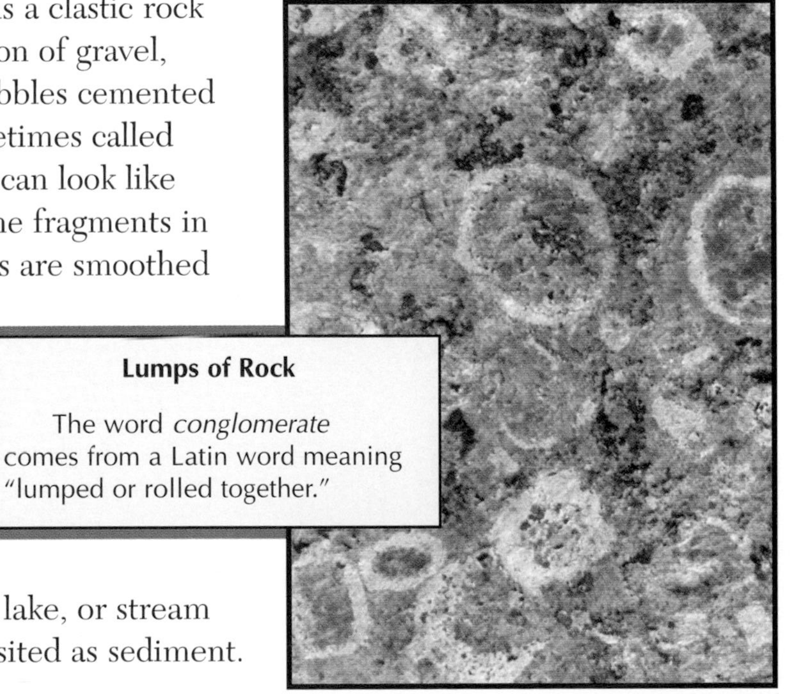

Conglomerate

Breccia is similar to conglomerate, but the fragments are not rounded. Breccia is gravel; sharper, angular cobbles; and pebbles cemented together. The rock pieces in breccia are sharper because they are not weathered as much by water as the conglomerate pieces are. Breccia fragments do not travel very far. They are usually found fairly close to the rock they originally came from. Flash floods are a source of breccia fragments. Breccia can also form from rock fragments that pile up at the base of mountains. These sharp rocks are cemented together over time.

The term *breccia* comes from an Italian word meaning "rubble."

Tumbling Rock Piles

The piles of rock that build up at the base of mountains are called *screes*.

Medium Grains

Sand grains measure .0025 to .08 inches across. The grains feel gritty and can be seen without a magnifying glass. Sand grains form the clastic rock called *sandstone*.

Sandstone forms in old deserts or in places where water was once flowing. This includes old riverbeds, **deltas**, lakes, or beaches. When water flows through the spaces between sand grains, the minerals in the water cement the sand together into rock. Thicker layers of sandstone are usually formed in deserts where wind blows the sand into very deep layers. Thinner layers are formed in old riverbeds or seabeds. Natural landforms, such as cliffs, ridges, and buttes, are often formed from sandstone.

There are different kinds of sandstone. The minerals present in each type determine its characteristics.

Sandstone can be soft or hard. It depends on how well the sand grains are cemented together. Sandstone cemented with calcite tends to be weaker as it is easily dissolved by water. Sandstone cemented by silica and iron oxide (rust) is much stronger.

Sandstone can have rounded or angular grains of sand. Round sand grains are usually formed in deserts and beaches. Angular sand grains are more likely to have been deposited on river bottoms.

The color of sandstone also varies from tan to brown to gray and even to red or pink. Graywacke is sandstone that is green or dark gray in color. Arkose sandstone is brownish gray or pink. Sandstone that contains iron oxide is a reddish color.

Sandstone arch carved by weathering and erosion

TRY THIS!

Make your own sandstone, and watch the effects of weathering. Mix ½ cup of plaster of Paris and 1 cup of sand in a large paper cup. Add ½ cup of water and stir. You may need to add a bit more water if the mixture is too thick to stir. Let the mixture dry until hardened. Use scissors to slit up the side of the cup to remove it. Place your sandstone rock in a safe spot outside where it will be exposed to weather conditions. Watch your rock for several weeks, and record any changes.

Small Grains

Silt and clay have the smallest grains. Silt grains range in size from .00015 to .0025 inches across. Grains of silt feel smooth when touched but gritty in the mouth. A magnifying glass is needed to see them. Siltstone is a gritty, brownish gray rock made of silt.

The Loess Hills of western Iowa were created when wind and rain moved silt deposits left behind by glaciers.

Windy Hills

Windblown silt forms a sediment called *loess* (pronounced "less" or "luss"). Large deposits of loess form hills that can support a variety of plants and wildlife. China has the world's largest and deepest loess deposit, estimated at almost 1000 feet deep.

Clay grains measure less than .00015 inches across. Grains of clay are so small that they feel smooth when touched or put in the mouth.

Clay is mostly made up of tiny particles of silica and alumina. Its color depends on which elements are also mixed in. Clay with a red color contains iron oxide. Clay with a gray or black color contains the element carbon.

Clay has many uses. Clay can be baked in an oven called a *kiln*. This baking process dries out the water inside the clay, bringing the tiny grains closer together. Bricks are made by baking the clay in rectangular molds. Pottery is made from clay molded into shapes. Cups, plates, bowls, pots, and vases are objects often made from clay pottery. White clay is called *kaolin* or *china clay*. It becomes porcelain when baked. Porcelain is used to make fine pottery, china dishes, and surfaces for sinks and tubs.

Shale is one of the most common sedimentary rocks. Shale is a clastic rock made of fine clay particles packed tightly together. Shale forms from ancient mud or clay deposits found in calm waters. The lower layers are pressed together by the weight of upper sedimentary layers. The water gets squeezed out, and the minerals cement the clay grains together. Because the grains are tiny, fine, and very close together, shale is a smooth rock.

Food for Thought

Elephants, gorillas, and macaws are among the animals known to eat clay. Scientists believe it helps calm their digestive systems.

Shale is formed in thin layers that split apart easily. Fossils are often trapped in these layers. Shale rock can be gray, reddish brown, or green.

Oil shale is a black rock containing a large amount of the element carbon. The carbon comes from dead animals and plants that are not fully **decayed**, or broken down. When heated, oil shale gives off a gas that forms oil when cooled.

Shale

A Few Facts About Carbon

- All living things need carbon.
- Air and water contain carbon.
- Carbon is present in the Sun and stars.
- Graphite (pencil lead) and diamonds are forms of carbon.

Sorting by Sediment Size

The size of the fragments within a clastic sedimentary rock can vary. Some rocks are well sorted or uniform. This means that all the particles within them are about the same size. Other rocks are poorly sorted. This means that they are made of a mixture of large and small particles.

Well-sorted rocks

Poorly sorted rocks

5

Organic Rocks

The Remains of the Dead

Organic means "related to **organisms**, or living things." Organic rocks are formed from things that were once living. When these organisms die, they break down, or decay. The remains of these organisms are then squeezed together into organic rock. Limestone and coal are organic rocks.

Lime Is Not Just a Fruit

Calcite is also known as calcium carbonate or lime. That's why the organic rock formed from calcite is called *limestone*.

Limestone

Limestone is made from the shells and skeletons of ocean organisms and the mineral calcite. Calcite is found in ocean water and is used by sea animals to build their shells. When the shell and skeleton pieces are buried on the ocean floor and pressed together over time, they form limestone.

Limestone is formed in shallow seas. Large deposits of limestone are often a sign that water once covered an area.

Limestone is usually white or yellowish. When mud becomes mixed in as limestone forms, the rock can be a gray color.

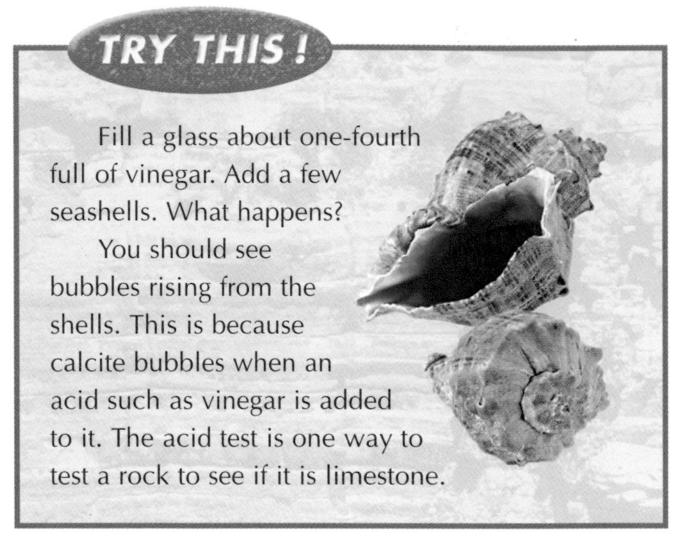

Fill a glass about one-fourth full of vinegar. Add a few seashells. What happens?

You should see bubbles rising from the shells. This is because calcite bubbles when an acid such as vinegar is added to it. The acid test is one way to test a rock to see if it is limestone.

Different types of limestone contain the remains of different organisms. Reef limestone is made of coral skeletons. Corals are sea organisms that have their shells on the outside. Coquina, or shelly limestone, is made from shell fragments of ocean animals. Freshwater limestone contains many snail shells. Chalky limestone is made from the skeletons of tiny sea organisms.

As you already learned on your trip to Carlsbad Caverns, caves are carved out of limestone. When rainwater dissolves large deposits of limestone, caves are formed. Inside the caves, calcite left behind from evaporating water creates spectacular cave formations, such as stalagmites, stalactites, draperies, and cave coral.

Writing with Rocks

The chalk you use to write on the blackboard is a type of limestone. This soft rock crumbles easily so it makes a good writing tool.

The actions of limestone and water create interesting land features. Clints and grikes are characteristics of limestone **pavement**. Clints are large blocks of limestone. Grikes are deep cracks between the clints. These cracks were carved by flowing surface water.

When water from rain and snow seeps through the soil and into a bed of limestone, it dissolves the rock, forming unique features. Areas covered with these special limestone formations are called *karst landscapes*.

Karst landscape in China

Sinkholes are common on karst landscapes. Sinkholes are cone- or bowl-shaped holes in the ground. These holes are made when a chunk of limestone is eaten away and cannot support the soil above. The soil then collapses into the sinkhole.

The passageways formed from water trickling down through cracks in limestone can eventually form tunnels and even caves. When surface water drains down into the passageways, streams and lakes can flow beneath karst landscapes.

Dolostone or Dolomite

Another organic sedimentary rock that forms karst landscapes is dolostone, also known as dolomite. Dolostone is very similar to limestone, but it contains the mineral dolomite instead of calcite. Many karst landscapes are a combination of limestone and dolostone.

Limestone

Coal

Coal is formed from decaying plants. This organic rock usually forms in swampy areas with many plants and not much oxygen. Layer after layer of dead plants, as well as sand and mud, pile on top of one another. The plants do not rot completely in the swamp water because the decaying process requires oxygen. The upper layers of partially decayed plant material push down on the lower layers. The water is squeezed out. Over time, different types of coal are formed.

Where Did All the Oxygen Go?

Swamps contain low amounts of oxygen because the water doesn't flow and mix with air. Bacteria and plants use up the oxygen originally dissolved in the water, and it isn't replaced with oxygen from the air. It's like a fish tank. Without a machine (aerator) to bubble air through the water in the tank, the fish would use up the oxygen and die.

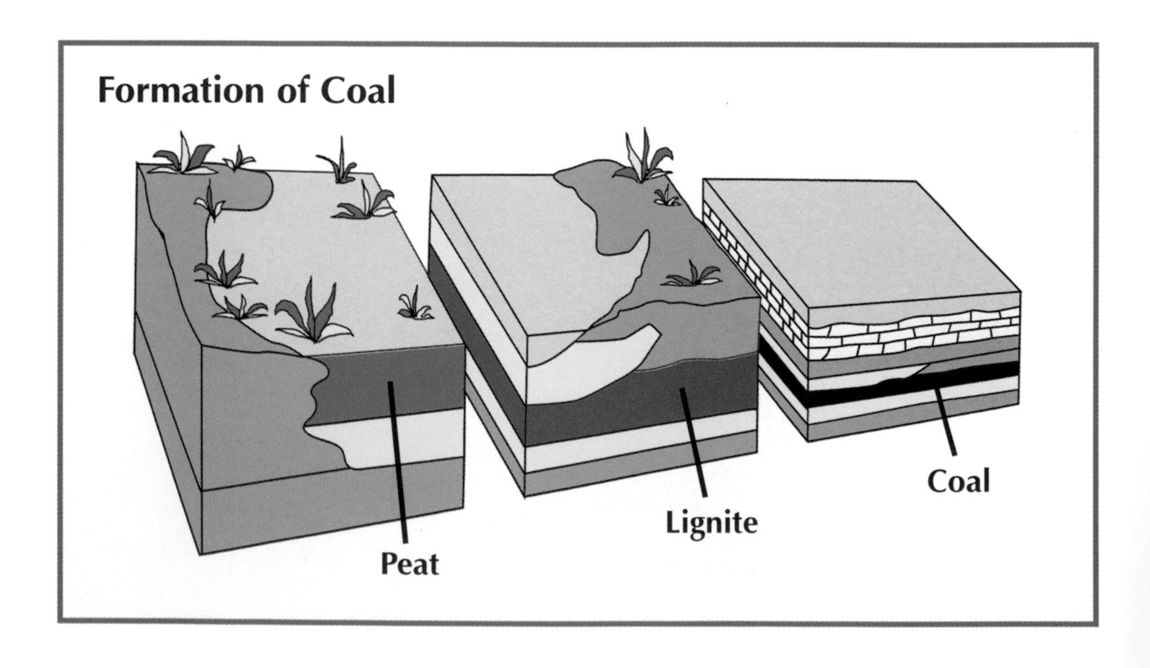

The quality of coal depends on how deep in the ground it is. Deeper coal layers contain more carbon and burn hotter. Peat is a soft, moist, slow-burning coal found closest to the surface. The next layer is lignite, a soft brown coal. You can still see evidence of decaying plants in these two coals.

Field of exposed peat

In the Wrong Book?

Anthracite coal is actually a metamorphic rock, *not* a sedimentary rock. Anthracite is formed at extremely high temperatures and under great pressure. It is the sedimentary rock coal changed by heat and temperature into the metamorphic rock anthracite.

Below lignite is bituminous coal. This coal is drier, harder, and blacker than the first two types. Found deep in the earth, anthracite is the highest-quality coal. It is more than 90 percent carbon. Anthracite is the most valuable coal but also the most rare.

Coal

6

Chemical Rocks

What's Left Behind

Rocks are made from minerals. Minerals are made from elements. Many of these elements are also chemicals. Chemicals are the basic substances that make up all things. Some single elements, such as hydrogen and nitrogen, are also considered chemicals. Elements can also be combined to form chemicals. A chemical reaction occurs when chemicals are mixed together and energy is absorbed or released as the chemicals change into new substances.

So what does all of this have to do with rocks? The third type of sedimentary rock is formed when minerals, or chemicals, dissolve in water and then settle out. These rocks are called *chemical rocks*.

What happens when you mix salt, sugar, or Kool-aid in a glass of water? The tiny solid particles dissolve in the water, forming a solution. All of the atoms and molecules (groups of atoms) that made up each small solid crystal are now evenly spread out between the water molecules. This also happens when certain minerals dissolve in water. When the water evaporates, the minerals are precipitated out, or left behind to reform crystals. Over time, layers of these minerals build up and their crystals become cemented together, forming chemical sedimentary rock.

Sedimentary rocks formed by water evaporation are also called *evaporites*. Evaporites often form in shallow waters or playa lakes. Playa lakes are dry areas where low spots are sometimes flooded. The water filling up the lake and then drying out over and over again leaves a thick layer of evaporite.

Dry playa lake

Halite

Halite, or rock salt, is an evaporite. Halite forms when salty seawater evaporates, leaving salt deposits behind. Over time, layers build up, forming beds of rock salt. Rock salt can be dug up, processed, and turned into table salt. Did you know you were sprinkling rocks on your food? You can also put rock salt on icy sidewalks to melt the ice and prevent accidents.

These salt flats are found in Death Valley National Park in California.

Halite is also found underground. Huge, thick layers of halite are mined by drilling into the layers, adding hot water to quickly dissolve the mineral, and then pumping the salty solution out. The water is then evaporated off, and the salt is processed.

TRY THIS!

Make your own salt bed. Stir together 1 cup of water and 4 tablespoons of salt in a glass bowl. Let the salty solution sit until all the water evaporates. (This may take weeks.) What's left in the bowl?

You should see square salt crystals on the bottom of the bowl and white frosty salt deposits on the sides of the bowl. The slow evaporation of the water from the bottom of the bowl left behind the salt crystals, or halite. The water rising up the sides of the bowl evaporated more quickly, leaving salt deposits behind. Crystals did not have time to form.

Gypsum

Gypsum is another evaporite made of the mineral sulfate. Water evaporating from old seas formed thick layers of gypsum. Gypsum is a very soft rock used to make plaster of Paris, casts, molds, and wallboards (drywall).

Gypsum

Limestone

Some limestones are organic rocks, but others are chemical rocks. Limestone can form when calcite precipitates, or separates, out of evaporating water.

Oolitic limestone is formed when calcite builds up on sand or silt grains underwater. These tiny balls of mineral and sediment are called *oolites*. The rounded oolites grow larger as they move back and forth in the water waves, picking up more layers of calcite. When oolites harden together, they form oolitic limestone.

A Fishy Rock

Oolitic limestone is sometimes called *roe limestone* because it looks like fish eggs (roe).

Chemical limestone can also be formed from existing limestone. Water traveling over or through limestone rock can wash some of the limestone away. The dissolved calcite in the water can later precipitate out. Layers build up over time, forming new limestone rock.

Travertine

Travertine is another chemical rock that forms when calcite accumulates. This often happens near hot **springs**. When the hot water bursts out of the ground, it carries calcite deposits with it. When the water evaporates, travertine formations build up. Travertine can also form along the banks of streams where the water splashes up and then evaporates. Travertine rock has holes because carbon dioxide bubbles are trapped in the rock as it forms.

Travertine formation

Polished travertine is similar to marble and often used for decorative purposes. Some flooring tiles and countertops are made from travertine.

7

Telling History Through Rocks

Geologists piece together the clues found in sedimentary rocks to help them determine the Earth's history. Since no people lived when the first rock layers were formed, there were no witnesses to early geological events. The layers in sedimentary rocks tell geologists interesting stories about the evolution of the Earth.

Today's scientists use various methods to figure out the Earth's timeline through rocks. They identify the type of rock in sedimentary layers, how thick the layers are, and the position of the layers. These facts help geologists piece together how the rock might have formed. Fossil evidence may provide information about the rock's age. By putting all the clues together, geologists can form an accurate picture of the Earth's geological history.

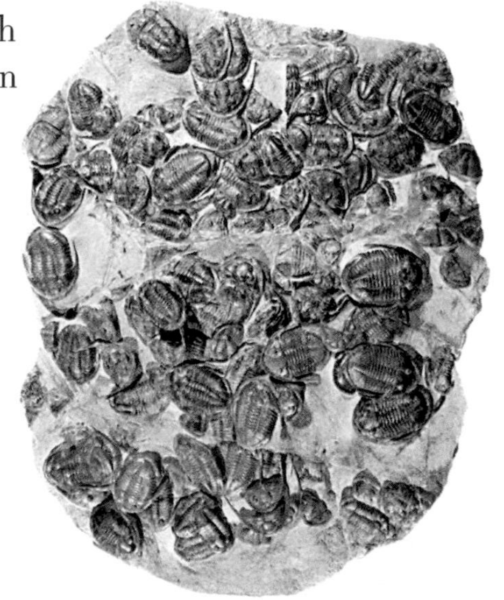

Trilobite fossils are found in rock that is 251 million to 545 million years old.

Early Discoveries

1600

Early scientists studied the Earth and formed conclusions that became the basis for the study of geology. In the mid-1600s, Nicolaus Steno came up with an idea called the Law of Superposition. This scientific law states that if sedimentary rock has not been disturbed, the older layers are on the bottom and the younger layers are on top.

Steno also used fossils trapped in the rock to identify the age of the layers. Since some plants and animals only existed during certain time periods, rocks containing those fossils must have formed during these periods. Species have also evolved, or changed, over time. Fossils showing these changes can be put in a time order based on their characteristics.

1800

In the early 1800s, British scientist William Smith agreed with Steno's theory that the age of rock layers could be figured out by looking at which fossils they contain. He concluded that completely different rock layers with the same types of fossils must have formed during the same time period.

1900

Absolute dating was first used in the early 1900s by Ernest Rutherford and B. B. Boltwood. This technique calculates the approximate age of a rock layer using the rock's chemical composition. Certain elements break down through nuclear decay. Each element decays at a different rate, or speed. By measuring the amount of different chemical elements present inside a rock layer, the age of the rock layer can be estimated.

Nuclear Decay

In some elements, the nucleus, or center, of an atom in the element can spontaneously turn into an atom of another element. This is called *nuclear decay.*

What's in a Layer?

Layers of shale

Sedimentary rock layers can show how the Earth has undergone major changes over time. Sedimentary rock known to have formed in or near water is often found in areas that are far from water now. This indicates that water covered the area at one time and some geological change has occurred over time.

Some sedimentary rocks, such as sandstone, shale, and limestone, that would have formed at or below sea level are now found at much higher elevations. This shows that the plates, or pieces, making up the Earth's crust have shifted over millions of years. Plates can uplift, fold, or even flip over, causing significant changes in sedimentary rock layers.

TRY THIS!

Fill an empty jar about three-quarters full of different-sized grains of sediment (sand, soil, small pebbles, medium-sized rocks, seashells, etc.). You can add the sediments in any order. Fill the rest of the jar with water. Put the cover on and shake hard for several minutes until the sediments are well mixed. Let the jar sit for many hours or overnight until the sediments have settled. What do you notice about the order of the sediment layers?

Due to the force of gravity, sediments are normally laid down in flat, horizontal beds. These flat beds are called *parallel beds*. In parallel beds, the sediments usually have time to settle in an organized pattern with the larger, heavier sediments sinking to the bottom while the smaller, lighter sediments remain on top. Rock beds with organized sediment are called *graded beds*.

If changes in the Earth's crust, the environment, or the climate occur during or after the formation of sedimentary rock, the layers may shift or tilt. Strong ocean waves, river currents, and desert winds can change the position of sediment layers. When the layers rest at angles instead of horizontally, it is called *cross bedding*.

Sedimentary cross beds

Unconformities

Unconformities are like missing puzzle pieces for geologists. Unconformities are changes in layers of sedimentary rock that occurred due to erosion, movement of the Earth's crust, or some other geological event. For example, if layers of rock are uplifted and then eroded away, there will be a gap between the old layers and any new layers that form on top. Sometimes one sequence of rock layers may lie flat on top of a very angled sequence. This indicates that the lower layers shifted due to movement of the Earth's crust and then new layers were laid down on top of the old ones. When unconformities occur, there is a gap in the time record of the rock. Sometimes millions of years of a rock's history are lost in a missing layer.

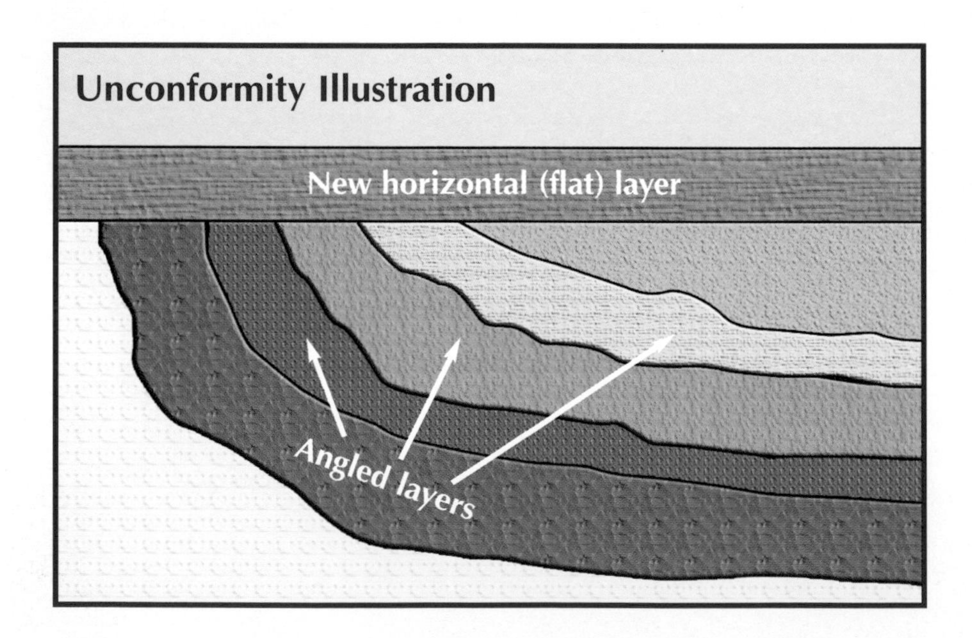

8

Sedimentary Rocks at Your Service

On your rockin' vacation, you came to recognize the beauty of sedimentary rocks. You might have "listened" to the geological stories these rocks have to tell. But what other uses do sedimentary rocks have? Throughout history, these rocks have provided valuable services for humans.

Early Uses

Long ago, humans used sedimentary rocks to make tools, weapons, and dyes. Flint is formed when the mineral silica settles out of water and forms lumps, or nodules, inside of chalky limestone. Flint can be chipped away to form very sharp cutting edges. Flint tools and weapons such as hand axes, spearheads, and arrowheads were used for hunting and protection.

Flint was also used to start fires. Hitting flint against other rocks creates a spark, which can be used to ignite grasses and tree branches.

Flint

Chert is a close relative of flint. Some ocean animals have lots of silica in their skeletons. This silica forms sediment layers on the ocean floor when the animals die. Over time, the silica layers are cemented together to form chert. Like flint, chert can also form as nodules inside limestone. When broken, chert's sharp edges made good spears and arrowheads.

Early humans also used sedimentary rocks, such as clays and chalk, for coloring. The rocks were crushed into powders and used as **pigments** in paints and dyes. These paints and dyes were used to decorate the body, clothing, baskets, pottery, jewelry, etc.

Glass

The use of sedimentary rocks to make glass was discovered thousands of years ago. Today, the "recipe" for glass hasn't changed much. It is still made using sand, **soda ash**, and limestone. Other minerals may be added for color. When these ingredients are melted at high temperatures, glass is created. Glass has hundreds of uses, including the manufacture of beads, vases, bowls, bottles, and jars.

Construction

Construction materials have been a common use of sedimentary rocks for thousands of years. The ancient Egyptian pyramids and Mayan temples were built from huge blocks of limestone. Clay has long been baked into bricks and pottery by many cultures.

Today's construction basics continue to be made using sedimentary rocks. Your home, school, and other buildings in your city or town were made with the help of these rocks. Limestone and sandstone are frequently used in buildings. Crushed limestone is an ingredient in **cement** and steel. Concrete is made of cement, sand, and gravel mixed with water. Gypsum is also used in the manufacture of cement as well in the plaster and drywall used inside buildings. Dolomite is used in road, bridge, and building construction.

Living in Brownstone

Brownstone is a reddish brown sandstone used to build houses in the eastern United States in the late 19th century. Many of these houses are still standing today even though the brownstone is deteriorating due to weathering.

Foods and Fertilizers

You already learned that your table salt comes from the sedimentary rock halite. But did you know that sedimentary rocks are used in other foods and in fertilizers added to soils that grow your foods? Phosphate ore is a sedimentary rock. The phosphate is used to make fertilizer, food **additives**, and the fizz in your cola. Gypsum is also used as a food additive in foods such as ice cream, spaghetti, and flour. Limestone is used to make the coating on many types of chewing gum.

Fuels

Imagine a cold winter morning with no heat or a day with no electricity to run your blow-dryer, radio, or computer. That would be a day without sedimentary rocks. These rocks provide several sources of fuel that give you heat, electricity, and gas for your car.

Coal, oil, and natural gas are fossil fuels formed in sedimentary rock layers. Fossil fuels are made from the remains of once-living things.

Coal is made from layers of dead plants that have been buried deep within the ground. The pressure cements these layers into coal. Coal is dug out of the ground and burned to release heat energy. At power plants, heat energy from coal is changed into electrical energy.

Oil and natural gas are made from tiny sea creatures. When they die, their bodies fall to the bottom of the sea. Over millions of years, layer upon layer of animal remains build up. Pressure from the upper layers turns the lower layers into oil and gas. The liquid oil and gas can travel up to the surface of the Earth through rocks such as sandstone and limestone that have large spaces between their particles. Oil and gas cannot flow through rocks such as shale that have smaller spaces between their particles. When gas and oil become trapped under a layer of shale, they form a pool. These pools can be drilled into so the gas and oil can be pumped up to the Earth's surface and burned to release energy. Oil can also be processed and turned into gasoline for cars, trucks, and other machinery.

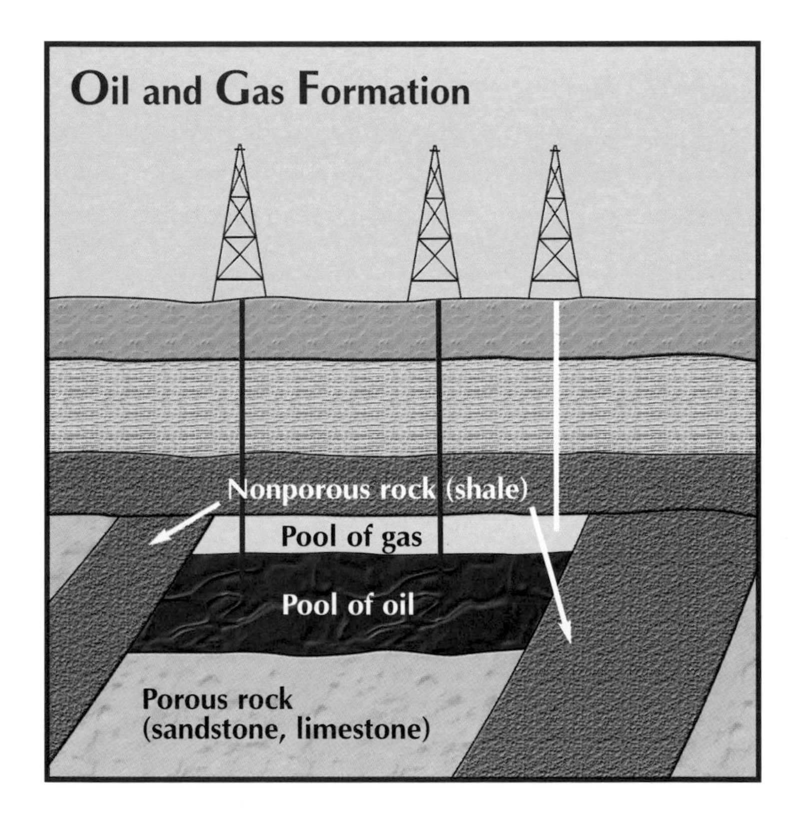

❖ ❖ ❖ ❖ ❖

From geological timelines to building materials and fuels, sedimentary rocks are invaluable. They keep us warm. They give us shelter. They make products that we use every day. Who knew that plain, old, ordinary rocks could be so important?

Internet Connections and Related Reading for Sedimentary Rocks

http://www.cobweb.net/~bug2/rock2.htm
Let the Rock Doctor tell you about the three types of sedimentary rocks and the ways they are formed.

http://www.minsocam.org/MSA/K12/rkcycle/sedimentary.html
This Mineralogical Society of America's "Minerals 4 Kids" site has a clear diagram of the sedimentary rock cycle as well as information about the formation and types of these rocks.

http://www.geocities.com/RainForest/Canopy/1080/sedimentary.htm
If you want to know more about a particular sedimentary rock, try this Web site. General information about the formation and properties of sedimentary rocks are followed by facts about specific sedimentary rocks.

http://core.ecu.edu/geology/harper/Sedimentary/Sedimentary.cfm
Looking for information on a sedimentary rock? Check out this index for a brief rundown on many of these rocks.

http://www.kaibab.org/geology/gc_geol.htm
If you want answers to your questions about the formation of the Grand Canyon, this site is the place to look. Lots of information and diagrams showing the different rock layers will teach you all you need to know.

Fossil by Paul D. Taylor. An Eyewitness Book on fossils. Dorling Kindersley, 1990. [RL 6.6 IL 5–9] (5865206 HB)

Rocks: Hard, Soft, Smooth, and Rough by Natalie M. Rosinsky. This book discusses the different types of rocks. Picture Window Books, 2003. [IL K–4] (3429306 HB)

•RL = Reading Level
•IL = Interest Level
Perfection Learning's catalog numbers are included for your ordering convenience. HB indicates hardback.

Glossary

additive (AD uh tiv) substance added to a food to improve or change the texture, flavor, color, etc.

atom (AT uhm) tiny particle that makes up everything in the world

calcite (KAL seyet) mineral formed when water evaporates from limestone (see separate entry for *evaporate*)

cement (SIM ent) to permanently stick together (verb); sedimentary rock ground into a fine powder (noun)

chemical (KEM uh kuhl) type of sedimentary rock formed when minerals settle out of water

clastic (KLAS tik) type of sedimentary rock formed from broken fragments of preexisting rocks, minerals, and shells (see separate entry for *preexisting*)

crystal (KRIS tuhl) solid form of a mineral with atoms arranged in a regular, repeated pattern (see separate entry for *atom*)

decayed (dee KAYD) broken down or rotted

delta (DEL tuh) deposit of sand and soil where a river meets an ocean (see separate entry for *deposit*)

deposit (dee PAH zit) to leave behind or drop off (verb); amount or quantity of sediment found in one spot (noun)

dissolve (di ZAWLV) to break down a solid in a liquid

element (EL uh ment) nonliving material made up of one type of atom (see separate entry for *atom*)

erosion (uh ROH zhuhn) movement of rock pieces by wind, water, or ice

evaporate (ee VAP or ayt) to change from a liquid to a gas

fossil (FAH suhl) hardened remains of a plant or animal

grain (grayn) small particle of sediment

mineral (MIN er uhl) nonliving substance made up of one or more elements (see separate entry for *element*)

organic (or GAN ik) type of sedimentary rock made from the remains of decayed animals and/or plants (see separate entry for *decayed*)

organism (OR guh niz uhm) living thing

pavement (PAYV ment) flat area of bare rock that resembles the surface of a sidewalk or road

pigment (PIG ment) substance that is added to give something color

precipitate (pruh SIP uh tayt) to become separated out of a solution through the evaporation of water (see separate entry for *evaporate*)

preexisiting (pree ig ZIS ting) already formed or made

reef (reef) chain of coral or rocks at or near the surface of water

rock (rahk) combination of two or more minerals (see separate entry for *mineral*)

sediment (SED uh ment) small pieces of rocks, minerals, shells, and soil

sedimentary (sed uh MEN tuh ree) type of rock formed when layers of sediment are pressed together (see separate entry for *sediment*)

soda ash (SOH duh ash) combination of the elements sodium and carbon used to make glass, soap, and paper

spring (spring) source of water coming from the ground

strata (STRAH tuh) layers of sedimentary rock

unconformity (uhn kuhn FOR muh tee) gap or change in layers of sedimentary rock

weathering (WETH er ing) natural process of breaking rocks down into smaller pieces

Index